Thoughts on Current Neuroscience

2nd Edition: Science and Health

Rowena Kong

First Printing: 2021

ISBN: 978-1-998518-43-2

Contents

Nicotine as a Muscle Relaxant

Nicotine, an alkaloid present in tobacco smoke, which is the main component responsible for its addiction, is an agonist of nicotinic acetylcholine receptors(nAChRs) which are present in the brain, the central nervous system and at both presynaptic and postsynaptic neuromuscular junctions of muscles. Nicotine activates nAChRs by mimicking the excitatory effect of endogenous neurotransmitter acetylcholine on muscles as it binds to these receptors, thereby stimulating muscular contractions. However, smokers also reported the opposing relaxation experience which suggested an additional physiological process involved. Webster (as cited in "More Effects of Cigarettes", 1969) observed short-lasting reduction in skeletal muscle tone of spastic patients who smoked, while Domino and von Baumgarten (1969) discovered the greater amount of nicotine smoked by their subjects, the greater depression of patella reflex they exhibited. This implicated both stimulant and depressant effects by nicotine are possible on the human skeletal motor system (Domino, 2001). Another study discovered the injection of nicotine into anaesthetised cats could produce depression of reflex responses by motoneurons after initially exciting sensory receptors in another region (Ginzel, Estavillo, & Eldred, 1975). The mechanism by which such depression worked could possibly be the powerful excitatory action of nicotine as a cholinomimetic on inhibitory Renshaw cells which are also stimulated by acetylcholine through the process of recurrent inhibition (Curtis & Ryall, 1966). Nicotinic-type receptors are found on these interneuronal cells which have synaptic connections with alpha motoneurons of the spinal cord that affect skeletal muscles (King & Ryall, 1981).

Domino, Kadoya and Matsuoka (1994) attempted to study the skeletal muscle relaxant effect of nicotine on smokers by measuring the subjects' Hoffmann reflex recovery cycles. Such recovery cycle provides a way to quantify the "excitability of the alpha-motoneuron pool in the spinal cord" which is involved in the ankle jerk (Domino, Kadoya, & Matsuoka, 1994, p. 527). Their hypotheses which predicted that smokers would have a depressed recovery cycle compared with non-smokers and that such decreased response would positively correlate with their blood plasma nicotine levels were confirmed. The authors stressed that nicotine has direct stimulation on the inhibitory effect by the Renshaw cells in the spinal cord as well as additionally exciting other peripheral chemoreceptors and sensory receptors that enhances the action of the Renshaw neurons on alpha motoneurons. In particular, nAChRs played a more significant role in the early phase of inhibition response as evident by the prolongment of this stage by higher nicotine levels in smokers. Another almost similar study conducted by Shefner, Berman and

Young (1993) examined the effect of nicotine on recurrent inhibition in the spinal cord via Renshaw cells. The results showed that for the 10 smoker subjects, there was a considerable plunge in their H-reflex response amplitude which suggested the activation of Renshaw cells. A study by Curtis and Ryall (1966) also discovered that nicotinic receptors on Renshaw cells are capable of rapid excitatory synaptic transmission.

Dourado and Sargent (2002) suggested that such synaptic transmission between alpha motoneurons and Renshaw cells was mediated by a certain α4β2 subunit of nicotinic cholinergic receptors on the latter in response to the former's stimulation. The authors confirmed from extracellular recordings that the activation of Renshaw cells by alpha motoneurons was nicotinic. According to Marubio et al., Picciotto et al., Whiting & Lindstrom and Zoli et al. (as cited in Dourado & Sargent, 2002), the α4β2-type receptors involved also provide high-affinity binding sites for nicotine. This brings to mind another idea by Marubio et al. and Picciotto et al. (as cited in Tapper et al., 2004) that α4β2 receptors are required for dopamine release by nicotine and this class of receptors could likely play a significant role in the dopaminergic reward system in the ventral tegmental area(VTA) and substantia nigra, leading to nicotine addiction. Pidoplichko, DeBiasi,

Williams & Dani (as cited in Tapper et al., 2004) supported from their study that nicotine increases the action potential firing frequency of dopaminergic neurons in the VTA which results in dopamine release in the striatum. In their study of mutant mice with their α4 receptors modified to be highly nicotine-sensitive, Tapper et al. (2004) found that low dose of nicotine was already capable of causing them to elicit dependency behaviour. Since smokers often spoke of nicotine's calming effect, regardless to curb their withdrawal symptoms, it is better not to rule out possible contributing factor(s) of the interaction of dopaminergic and GABAergic neurons, on which nicotinic cholinergic receptors are also present and active, towards evoking the physiologic relaxation nicotine often offers (Giniatullin, Nistri, & Yakel, 2005; Silverstein, 1982).

Wall Shear Stress and its Causality in Cerebral Aneurysm Development

In a recent case-control study by Zimny et al. (2021), the development of unruptured cerebral artery aneurysm can be reasonably traced to the combined factor of high wall shear stress and wall shear stress gradient. In addition, regions of bifurcation apices with such factor of significant wall shear stress are critical in aneurysm formation. A consideration of directional and spatial changes in opposition against the inertia and seamless hemodynamic blood flow in approaching the juncture of bifurcation apex that increase interaction of contact forces between arterial walls and blood volume may aid in further discussion of causality. On the other hand, arterial diameter or radii is also a contributing factor based on the observation of smaller vessels in female patients at bifurcation points that lead to high wall shear stress due to greater concentration of force within a more compact space from increased blood flow velocity (Lindekleiv et al., 2010). A probable explanation for the increased tendency of aneurysm development in such regions could also look into the opposing reactive mechanism of vessel walls to expand their surface area and volume in order to counter, accommodate, and redistribute the increased amount of force velocity from instances of more aggressive blood flow and hemodynamic contact, even of a transient nature that builds up over time. This suggests that an interplay of blood flow and fluid force mechanics and the physical characteristics of arterial curvature has a dynamic role to play in the localization of cerebral aneurysm formation.

Hyperthermia-Induced Vasoconstriction: A Physiological Counter Mechanism

Hyperthermia-induced vasoconstriction is a paradoxical physiological phenomenon that has limited explanation and exploration in research, despite being observed in clinical case studies. The bodily response of vasoconstriction and vasodilation to increased heat loss and gain respectively work on the principle that heat transfer through the amount of vascular surface area exposed can be moderated in order to be minimised or maximised. Hyperthermia-induced vasoconstriction likely transitions from the break point of maximal limit of vasodilation to counter extreme heat loss beyond this point and to mitigate the potential detrimental effect in arterial pressure disturbance by extreme temperature condition. We also propose an additional explanation of this physiological counter mechanism based on the perspective of body core-peripheral steep temperature gradient reduction during hyperthermia or significant temperature increase, and a spontaneous tendency to avoid unstable blood pressure perturbation.

The paradoxical phenomena of hyperthermia-induced vasoconstriction, hyperthermia-induced hypothermia, and hyperthermia induced arterial vasoconstriction have been observed in animal heatstroke clinical case studies (Romanovsky & Blatteis, 1996). We attempt to explain the basic mechanism behind such findings in terms of the principle of body heat conservation and natural countermeasure against heat exposure through reduction in the extent of vascular surface area exposed. One study which examined the outcome of heating a rabbit's carotid artery produced graded vasoconstriction which is proportional to temperature increase (Mustafa, Thulesius, & Ismael, 2004). The bodily response of vasoconstriction and vasodilation to increased heat loss and gain, works on the principle that heat transfer through the amount of vascular surface area exposed can be moderated in order to be minimised or maximised through muscle activation, which in turn also results in volume changes.

Regulating Thermodynamic Temperature-Volume-Pressure Disturbance

Logic explanation would support persistent physiological response of vasodilation following hyperthermia, yet observations reported that the opposing response of vasoconstriction has taken place. We propose an explanation that under the condition of extreme and enduring hyperthermia, continuous vasodilation would be health-compromising on the system and a critical factor of hypotension or severe arterial pressure reduction would result should a break point of maximal limit on vasodilation be not set in place to counter further rapid heat loss beyond this point. Vasodilation inevitably has an effect on vascular pressure because smooth muscle relaxation and expansion of the volume of blood arteries and vessels increase space and reduce pressure, an interplay of thermodynamic equilibrium. Hyperthermia-induced vasoconstriction prevents the detrimental effect of continuity in progression of vasodilation in relation with vascular pressure, and functions to maintain internal pressure gradient's constancy and stability, as constricting blood vessels would elevate arterial pressure and perfusion of core body organs and system, such as the experimental rabbit's carotid artery in a previous research. In order to keep the arterial pressure constant and unperturbed, reduction in volume through vasoconstriction is needed to counter the hypotensive effect of sharp temperature increase and corresponding vascular volume expansion(dilation) of hyperthermia. In the case of the experimental rabbit's carotid artery, graded temperature increases with decreasing volume through vasoconstriction, internal arterial pressure would remain constant and be preserved to maintain thermodynamic equilibrium. In addition, heat loss to the extreme through exposed surface area is also regulated and moderated. In addition, we include another explanation of this physiological counter mechanism based on the perspective of body core-peripheral steep temperature gradient reduction during hyperthermia or extreme temperature condition and a spontaneous tendency to avoid unstable blood pressure perturbation. Such spontaneous physiological counteractive measure to minimise vascular surface area and volume of heat exposure through extreme vasodilation by inducing vasoconstriction to counter against persistent thermodynamic disturbance of internal organ(s) strategically restores and re-elevates potential significant arterial blood pressure reduction.

Moderation of Vascular Surface Exposure

Experimental results pointed out a spontaneous physiological counteractive measure to minimise organ and/or arterial surface area heat exposure through vasoconstriction which might otherwise prompt excessive heat gain/loss from the ambient environment, and/or vasodilation, should it occur. In the event of vascular surface exposure, there may be a bi-directional heat flow exchange that can be lost or gained based on inter-surface concentration gradient through the blood vessel barrier. When a blood vessel or artery constricts through smooth muscle activation, the amount of vascular surface area being exposed, whether to heat that is external from or internal within the body, is reduced. Since such vasoconstriction is proportional to temperature increase, the speed or rate at which this constriction responds and therefore, the extent of vascular surface area being reduced, may be drastic with extreme means of bodily cooling under a heat wave environment, such as whole-body cold-water bath or shower. The reason being extreme vasoconstriction of both cerebral and peripheral blood vessels, which may result in increased intravascular pressure due to intravascular spatial volume decrease and as an above study concluded, cerebral blood flow decrease, and ischemia brain damage. On the other hand, cold-induced vasodilation is the dilation of peripheral arteries under exposure to extravascular cold temperature (Daanen, 2003; Flouris, Westwood, Mekjavic, & Cheung, 2008). It is counterintuitive but the build-up of a heat concentration gradient with greater intravascular heat flow than the external may prompt active heat loss and exchange across the vascular surface into the peripheral environment. Another factor to consider is the body's spontaneous counter-reaction and opposition against cold-induced increase of intravascular pressure to occur under extreme low ambient temperature, which then triggers the vasodilatory effect as a result to prevent excessive internal pressure rise within the arteries. It is no wonder that cold-induced vasodilation helps to reduce cold injuries, which may be attributed to its counter mechanism strategy to decrease severe intravascular pressure increase under extreme cold conditions (Kingma, Hofman, & Daanen, 2019).

Body Core-Peripheral Temperature Gradient

Although cold-induced cutaneous vasoconstriction in humans bears functional significance in heat and blood flow reduction to external environment through contact surface area exposure of the skin blood vessels, hyperthermia-induced vasoconstriction of the animal's carotid artery is better viewed from the perspective of core body temperature (Alba, Castellani, & Charkoudian, 2019). In the event of hyperthermia-induced vasodilation of cutaneous vessels, a temperature gradient with the internal arteries of the body core is induced. In order to counter the increasing temperature gradient and maintain homeostatic equilibrium, cutaneous heat loss needs to be matched with core heat gain. In addition to the presence of a protective blood-brain barrier, the carotid artery has to preserve a strict boundary for the brain from extended cutaneous heat loss and the attempt of core body heat loss to lessen the temperature gradient with peripheral blood vessels.

Preservation of Tissue Perfusion

A large body core-peripheral temperature gradient is related to poor tissue perfusion and decreased vascular function as a narrow range of core-peripheral difference of 4° C or less is indicative of normal circulation (Schey, Williams, & Bucknall, 2010). Hence, paradoxical, yet protective counter mechanism of vasoconstriction of the carotid artery may be necessary to preserve vulnerable brain components from probable associative heat loss during the event of hyperthermia and significant cutaneous vasodilation at the cost of extended core-peripheral temperature gradient. Therefore, the body's heat conservation strategy remains in effect and as previously noted, reduced vascular surface area exposure helps maintain a fairly optimal body core-peripheral temperature gradient to prevent unsteady blood pressure perturbation beyond the break point of hyperthermia-induced vasodilation transitioning into vasoconstriction. The three fundamental thermodynamic states of vascular temperature, pressure and volume work closely and persistently to maintain equilibrium and stable gradients within a system where the impact of changes on one induces a unitary response by all states as a counter mechanism.

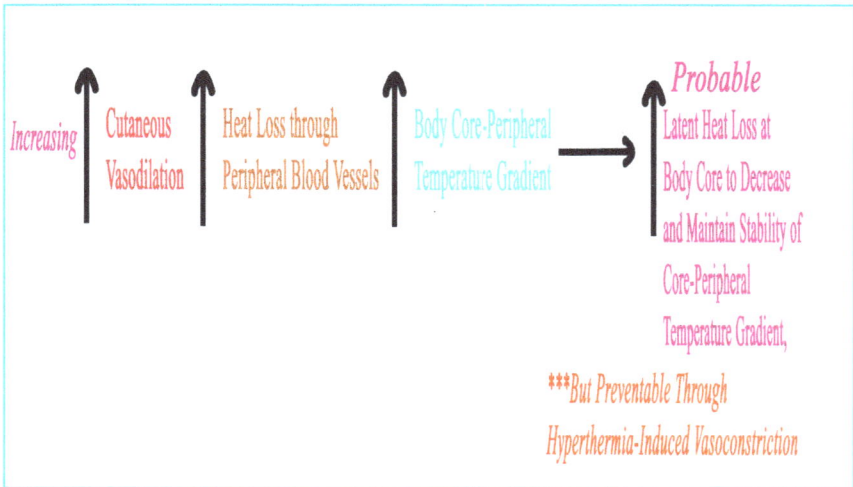

Figure 1: Vasoconstriction Counter Mechanism Hypothesis Thought Flow

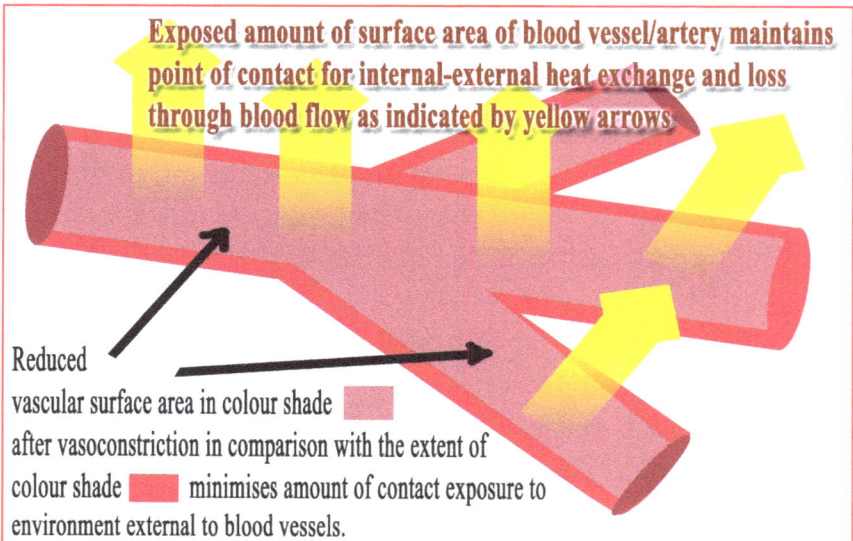

Figure 2: Vascular Surface Area Adaptability to Heat Flow Exchange Variability

Nevertheless, this does not preclude a likelihood of dysfunction as the body adapts to extreme environmental temperature disturbance(s) that alter physiological balance and homeostatic stability. Measures to cool the ambient environment and atmosphere may be necessary before direct and/or invasive means of patient treatment to minimise progression to extreme stages of signs and symptoms.

Anosognosia: How Much Do We Know?

Anosognosia, a term used to describe the lack of awareness of a patient regarding his/her dysfunction or medical condition, was first coined by a neurologist called Joseph Babinski more than a century ago. Now, this disorder or deficit is being studied in research across health disciplines with the goal of developing feasible interventions to minimise its impact on the process of recovery for patients with this disorder. One of the tests a researcher or healthcare provider would use to diagnose such disorder in patients would be to ask certain questions like the choice of using either the impaired hand to work on a task as any normal person would or just the other healthy functioning hand in an easier task. A very simple straightforward question indeed, but one which makes me wonder about the chance of this question gauging the degree of denial of the patient's impairment rather than a real "un-awareness" of it. On the other hand, could there be a subtle component of self-denial involved in this presumption of deficit of self-awareness in the patient that results from the disproportionate hopefulness and optimism for an immediate full recovery or blurred recollection of recent injury which produces the impairment? Apparently, the psychological and mental state of the anosognosic patient is a critical target for intervention should we need to gain a more in-depth and individual-ised understanding of this by-product of the overall condition. Self-denial can be purely psychological while awareness is in part a neurological function. In order to identify and isolate psychological denial from self-aware-ness, personalised attention that caters to and probes the unique needs of each anosognosic patient is required be-cause injury, dysfunction and pain are a very different and personal experience for everyone, and no two cases are exactly alike. The physical manifestation and outcome of cases of dysfunction may be easily grouped under sim-ilar categories, but because every individual is unique and different from everyone else, the psychological impact along with expectations held for prognosis by each pa-tient is remarkably varied and very personal. When the disorder of anosognosia is being described as a deficit of self-awareness, questions arise as to the extent, boundary and nature of the deficit. Is it a total or partial lack of knowledge of the impairment per se or that it includes other intact function(s) of the impaired body part? How is the patient's memory content about the events pre- and post-injury? Is the patient able to direct attention to the impaired body part spatially and temporally in a normal manner? Such questions may offer clues to signs of psy-chological abnormality resulting from the physical impairment.

Looking from an alternative angle, if we are to rule out self-denial and unrealistic optimism from lack of self-awareness, there is a possibility that the component of awareness or consciousness of a particular brain function is proximally located within the same brain region(in such cases, affected by an injury or infarct) responsible for that certain function, e.g. voluntary muscle movement, sensory impulse, etc. In other words, awareness/consciousness of specific brain functions could be localised or compartmentalised and spread across various regions of the brain, much like memory, which is now no longer considered a unitary construct. It could be one thing to be seeing a flower and another thing to be actually aware that you are seeing it at this particular instant. This extends to the modern understanding of human consciousness as being built upon by an integration of multiple, yet very specific, networks spanning various regions of the brain, and that our attention to a particular theme or subject at any one instance, is but a very limited and one of a myriad of processes happening in parallel within the most complex organ of the human body. Like a stage performance, the actors of psychology, cognition, emotion and neurology play their individual roles but they all add up in concert to the consolidated screenplay, though without rehearsals beforehand but having everything taking place in real-time, a miss or mistake of a line

or two of the script may throw in some dysfunction or impairment along the way. Then, the rest of the actors
would have to adapt and play along to the error...

Towards a Neuropathological Definition of Psychiatric Disorders

Perhaps, by now, people have begun talking about the genetic connection between autism and psychiatric disorders of schizophrenia and bipolar disorder. At first thought, it could be presumed that these gene variants may be involved in some way in coding for and building up the protein components that form a part of the neurons in our brain, which in turn effect the health and function(s) of the neurons. A geneticist could do better in going deep into the details of which of these protein components these gene variants map unto, but most general learners could also guess the effect of these disheveled proteins can simply be negative in such a way that it spreads over and compromises the normal nervous functioning of an individual. Thinking along this line, one could go on to picture that the best way to get a clue of the genetic underpinnings of this newly discovered connection between such disorders is to compare and contrast the symptoms presented in autism, schizophrenia and bipolar disorder, e.g. impaired cognitive and social functioning(autism and schizophrenia), blunted affect(autism and schizophrenia), mania(bipolar disorder and schizophrenia), etc. Where there are positive correlations between the disorders, it may suggest an underlying genetic interaction after shared environmental variables have been ruled out.

In response to the motivation to study such mosaic of genetic connections, it is now known that this relationship is built upon gene expression which affects the supportive microglia cells that have a role to play in the immune function in our brain. This aside, one could wonder if we have for a long time limited our understanding and exploration of autism, schizophrenia and bipolar disorder by their discrete categorisation in the DSM-5 in terms of their bounded distinction from any other disorder. We are increasingly seeing overlap in symptoms between disorders which are presumed to be unrelated to each other. A psychiatrist who has just diagnosed a patient with having autism spectrum disorder may be unlikely to give a second thought about secondary symptoms which could be suggestive of a concurrent psychosis. The point is, we are still in the early stage of uncovering whichever relevant brain regions are specifically involved in certain disorders, yet it is becoming clear that there is tremendous overlap due to the highly compact nature of neuronal distribution in our brain. When the pathological pattern involves similar brain regions, it gives the explanation why there is symptom overlap, which is why we need to expand the scope of the disorder and redefine the boundaries...This would suggest a limitation of diagnostic criteria that is based solely on the DSM-5...It seems that it is insufficient for us to diagnose a disorder in terms of manifestation of overt psychiatric symptoms alone...We would seldom encounter a patient's complaint of a fixed single symptom all by itself but it is rather more of a group combination. Perhaps, covert neuropathological evidence is a more comprehensive supplementation to mere patient verbal assessment and observation which is prone to different healthcare providers' biases and subjective judgements. In taking the anatomical and neuroscientific aspects of brain pathology into account and linking them with interconnected symptoms at the expense of the blurring of disorder criteria boundary, we may be one step closer to accuracy and productive management of a condition.

Neurogenesis in the Case of Recovery from Vegetative State

The process of adult neurogenesis has long been widely popularised as the birth of new neurons in the mammalian brain since its discovery more than 50 years ago by Joseph Altman (Altman, 1962). Although research challenges and controversy are still present, studies are increasingly exploring such a promising working of the resilient human brain (Lee & Thuret, 2018). In psychiatric and neurodegenerative disorders, it is only to be expected that illness progression would compromise the brain's ability to undergo neurogenesis at the same rate as healthy individuals (Apple, Fonseca, & Kokovay, 2017). However, there may even be more challenges facing the study of neurogenesis in patients who suffer from traumatic brain injury. Perhaps, an autopsy could not be more useful for such a purpose in providing aiding evidence to structural imaging findings. Russian researchers Vainshenker, Zinserling, Korotkov and Medvedev (2017) reported the case of a patient's recovery from vegetative state following a severe brain injury with incidences of skull fracture and subarachnoid haemorrhage. After a period of PET scan procedures and observation of functional improvement in neurologic symptoms, e.g. consciousness level, motor, visual and pain responses, the female patient died from cardiac arrest and a histologic examination was performed on her brain.

The PET scans which the patient underwent during her period of recovery positively indicate increased glucose metabolism in a variety of regions - the frontal, temporal, occipital lobes, the brainstem and cerebellum. Does such an increase in distribution of energy resources and utilization correspond to possible new birth of neurons and network connections that might be simultaneously taking place to spur recovery processes? Well, the outcome of the histologic examination shows remarkable discovery of neural stem and progenitor cells (Musashi1+, Nestin+, PCNA+, and Ki67+ cells) in widespread areas of the hippocampus, frontal, parietal and occipital cortex, caudate, thalamus, mammillary bodies, brainstem, cerebellum, and near the posterior horn of the lateral ventricle. What is obviously interesting from this report is that such new neural stem and progenitor cells are present in regions outside neurogenic areas of the the subgranular zone in the dentate gyrus of the hippocampus and the subventricular zone (SVZ) of the lateral ventricles. Whether severe neural damage provide elevated stimulation for such on-site growth or cell migration from neurogenic areas remains to be questioned and further investigated. Nevertheless, there could be a reason to believe the subtle likelihood for either or both to happen and it does challenge certain long-standing pessimistic scientific views that impose limitations on the prospect of neurogenesis in the adult human brain. In this case report, the association of neurogenesis with brain injury recovery cannot be ruled out in light of the evidence, albeit modest and unique...

Caffeine and Memory Consolidation

We all know coffee, with its caffeine content, that it is an effective stimulant which increases our arousal and keeps us awake while we pull our all nighter during grueling finals (and midterms for tougher courses). It is 'addictively useful' for both exams and socializing, regardless of whether one is a university student or not, and not surprisingly, researchers are also baffled by its popularity that they could not help but attempt to dig deeper under its superficial appeal in search of possible hidden benefits for our health. And it was not without success, for progress has been made in reports of caffeine's associations with neuroprotective effects against Parkinson's disease, Alzheimer's disease, and even potentially beneficial against chronic and fatty liver diseases. These benefits are still under ongoing study, and while we do appreciate the long-term value of such discoveries, as students, we are likely more concerned about how coffee can impact us academic-wise.

It turns out that caffeine research in the area of cognition and attention is not a new thing, though results obtained have been mixed. Depending on the condition under which learning takes place, caffeine can be a mild cognitive enhancer due to its cumulative influence on arousal, task performance and concentration, among other factors. Still, that does not deter students from consuming it. I, for one, do so on a regular basis, for the lame excuse of mood improvement. However, I have also found another reason for not quitting, and that is after reading a recent research report of how caffeine may even help promote memory consolidation, a process which converts information encoded in short-term memory into long-term storage (Borota et al., 2014; Kelly, Mikell, & McKhann, 2014). A group of researchers at Johns Hopkins and University of California, Irvine, led by Dr. Michael Yassa, tested study participants by administering a surprise memory test a day after they were shown a set of over a hundred pictures and given a pill which contained either caffeine or a placebo. Unlike some previous studies, participants took the pill after they were acquainted with the pictures, for the purpose of testing the effect on memory consolidation and the results did show that participants who consumed moderate doses of caffeine were better able to discriminate between the first and a new set of pictures shown the day after. In other words, these participants were able to tell more accurately than those who received the placebo that certain items in the new picture set were only 'similar' instead of 'old' or have already appeared in the previous set. Somehow, this helps shed a positive light on the potential memory-enhancing effects of caffeine.

While research has looked into the positive effects of caffeine on working memory, people may still be skeptical of its unclear and likely mechanisms, though most focused on neurotransmitters adenosine(involved in wakefulness) and dopamine(involved in pleasure and reward learning) receptors, and how it can be generalized to more complex long-term memory beyond simple object recognition tasks. There is also the implication of a role of dopamine in memory processing which is emphasised less in the literature, since caffeine tends to act indirectly on dopamine receptors through its antagonism on adenosine receptors. But just in case you are wondering if you should be drinking coffee during or after regular, unrushed quiet study times rather than only during highly stressful final weeks(even though caffeine has been shown to lessen the negative effects of stressors and sleep deprivation and improve cognitive performance at the same time), I hope the above research could provide you with enough incentive to experiment and experience the personal results for yourself. After all, every student should be developing healthy study habits when aiming for success, whether she has a fondness for coffee or not.

Early Onset of Schizophrenia in Men

Research has noted that there are sex differences in illness onset, outcome and the type of symptoms experienced by schizophrenic males and females. Studies have shown that schizophrenia began at an earlier age for men (Evenson, Meier, & Hagan, 1993; Goldstein, Tsuang, & Faraone, 1989; Troisi, Pasini, & Spalletta, 2001). As a result of their earlier onset, men show less developed premorbid functioning with poorer treatment outcome, such as persistent social withdrawal and speech disturbances (Salokangas & Stengard, 1990). Men also experience more severe negative symptoms and display more structural brain abnormalities (Salokangas & Stengard, 1990; Troisi et al., 2001).

Since structural brain abnormalities are significant in men, research has investigated the feature of hypofrontality. In one study, the MRI scans of schizophrenic males demonstrated hypofrontality particularly in decreased metabolic activity of the dorsolateral prefrontal cortex compared with non-schizophrenics (Molina et al., 2005). In this case, the drug treatment given did not help the subjects' negative symptoms. Another study discovered lowered whole brain volume, cortical and temporal lobe gray matter volume, noticeably the left temporal lobe, which is exhibited in the schizophrenics' poor language and verbal ability, and their significantly greater lateral and third ventricular volumes (Fannon et al., 2000). With their focus on the relationship between frontal size and symptoms, Andreasen et al. (1986) observed smaller frontal areas of schizophrenics in about 40% of the male subjects, which are likely due to reduced dorsolateral and orbital regions. Despite the weak association between low frontal sizes and negative symptoms, they found that smaller cerebral and cranial sizes were strongly related to such symptoms. Perhaps the relationship between hypofrontality and negative symptoms cannot be more strongly emphasised than in the comparison of these symptoms with those of behavioural variant frontotemporal dementia(bv-FTD) (Ziauddeen, Dibben, Kipps, Hodges, & McKenna, 2011). Based on the proposal that "negative symptoms are a manifestation of impaired frontal lobe function", a study produced similar pattern of scores between the schizophrenic and bv-FTD subjects on measurements of negative symptoms. Schizophrenic symptoms like affective flattening and avolition-apathy were scored by bv-FTD subjects which might closely relate with their emotional inexpressiveness. Schizophrenics also scored on bv-FTD's Frontal Systems Behaviour Scale for apathy. Speech deficits shared between both groups could be possibly attributed to reduced frontal lobe function. As for sex differences in brain processes, one MRI study looked into the cerebral gray matter volumes of healthy young girls and boys during their period of brain maturation of synaptic pruning. The rate of decrease in gray matter volume in boys is faster than that of girls (De Bellis et al., 2001). An explanation is that estradiol in girls could have delayed such pruning process and played a part in neurodevelopment with its receptor distribution in the brain.

While the differences in severity of illness are more pronounced between young men and women, such contrast is not for post-menopausal women. In fact, research has suggested the protective effect of oestrogen for women since the late 70s (Häfner, 2003). Fink (as cited in Häfner, 2003) found out from animal experiments that the stimulation of the serotonin transporter gene by oestrogen has a protective, antipsychotic function similar to decreased D2-receptor sensitivity. Such receptor density was reported by Kaasinen et al. (as cited in Rao & Kölsch, 2003) to be higher in post-menopausal women. Riecher-Rossler et al. (as cited in Häfner, 2003) discovered increasing oestrogen plasma levels correlated negatively with schizophrenic symptom scores of women with regular menstrual cycles. Oestrogen also appears to enhance the effectiveness of neuroleptics taken by schizophrenic women. Frisch and Gur et al. (as cited in Rao & Kölsch, 2003) reported that increased cerebral blood flow and more body fat in women also aid in more efficient bodily distribution of lipophilic antipsychotic medication. Pozzo-Miller et al. (as cited in Rao & Kölsch , 2003) mentioned that a female sex hormone, estradiol - 17β, which indirectly prevents neuronal cell loss by controlling intracellular calcium concentrations, is presumed to interact with NMDA and AMPA receptors. Javitt (as cited in Rao & Kölsch, 2003) stated that drugs which reduce negative symptoms also work by enhancing these receptors' functions.

Brain atrophy and low oestrogen level aside, studies have shown less promising pharmacological treatment of negative symptoms which are more prevalent with men despite hopes with second-generation antipsychotics(SGA) (Erhart, Marder, & Carpenter, 2006). Adjunct treatments do not have well-grounded efficacy and are not widely used. Overall, the effectiveness of SGAs and available adjunct medication in treating negative symptoms is questionable and explains poorer prognosis of schizophrenic men.

Immunomodulator Adenosine and Energy-Conserving Strategy

Adenosine is known as a neuromodulator with a broad range of functions across neurophysiological systems in cerebral, vascular, renal, immunological and muscular health, to name a few. Intracellular and extracellular adenosine can be generated by several sources and pathways, but more commonly through phosphohydrolysis of the high energy stores of extracellular adenosine triphosphate and adenosine diphosphate (ATP/ADP) by enzymes CD39 (nucleoside triphosphate dephosphorylase) and CD73 (ecto-5'-nucleotidase). There is ample research documenting possible neuroprotective mechanisms performed by adenosine through its receptor activation that sets into motion cascading processes which aid in promoting and maintaining neural and immune health (Alam, Costales, & Cavanaugh et al., 2015; Bauerle et al., 2011,; Deaglio et al., 2007). However, lesser known and discussed is the implicit energy store regulating role performed by adenosine through various channels and receptor transmission that can be both simply understood and complex. Based on the close connection of adenosine with its origin precursors of the family of adenosine phosphates (ATP/ADP/Adenosine monophosphate [AMP]), it could be proposed a fundamental strategy of energy conservation mechanism by adenosine through a process of interactive recycling and replenishing that persists in maintaining a consistent storehouse of ATP-sourced energy which caters for a host of balanced neurophysiological functions. In a sense, adenosine is a significant and crucial player as a relaying mid-point between its precursors and reconsolidated phosphate variant products such that the continuous chain of energy supply can be maintained and regulated in an efficient and uninterrupted manner under normal homeostatic surveillance of neurophysiological conditions.

In this discussion, an energy conservation strategy proposed to apply to the relevant neurochemical elements, is an implicit mechanism that helps preserve a regulated flow of energy production maintained through various efficiently orchestrated neurochemical reactions that source, exchange and reform precursor and product molecular units with the goal of minimising excess energy expenditure and maximising reusability. Under such proposed definition, an example can be demonstrated by the involvement of adenosine in dampening the process of inflammation through dephosphorylation of ATP, which increases during tissue inflammation, by enzymes CD39 and CD73 (Alam, Costales, & Cavanaugh et al., 2015;

Deaglio et al., 2007). Such mechanism implies a pathway of depletion of extracellular ATP which is consequently being converted to adenosine as the end product to discontinue the process of inflammation that could potentially further tax cellular energy stores and harm the human host of a pathogenic invasion. It can be inferred that ATP is a necessary resource supply for the performance of the immune response task of inflammation and that the halting of such work by its conversion to adenosine therefore helps set the brake on the process, be it of immediate necessity or leading to a prospective long-term significance based on an individual's physiological capacity to provide and endure, thereby strengthening the concept of energy-conserving role played by adenosine. Nevertheless, such strategy does not preclude the reformation of ATP from adenosine at other potential sites of need after a temporal gap in order to maintain a steady flow of the energy pool through various channels.

In conclusion, the human physiological system is an integrative and autonomous entity that is almost entirely energy resource-dependent based on consideration of the needs and demands for sustenance of multiple make-ups of such a system that works in a complex fashion of intensive and orchestrated cooperation. Through such careful perspective and analysis of energy provision and needs equation, it is hoped that scientific endeavours that aim to better our understanding of human physiology could build more robust and sound component principles that lead to promising and evidence-based results.

References

Alam, M. S., Costales, M. G., Cavanaugh, C., & Williams, K. (2015). Extracellular adenosine generation in the regulation of pro-inflammatory responses and pathogen colonization. *Biomolecules, 5*(2), 775-792.

Alba, B. K., Castellani, J. W., & Charkoudian, N. (2019). Cold-induced cutaneous vasoconstriction in humans: Function, dysfunction and the distinctly counterproductive. *Experimental physiology, 104*(8), 1202-1214.

Altman, J. (1962). Are new neurons formed in the brains of adult mammals? Science, 135(3509), 1127-1128.

Andreasen, N., Nasrallah, H. A., Dunn, V., Olson, S. C., Grove, W. M., Ehrhardt, J. C., … Crossett, J. H. W. (1986). Structural abnormalities in the frontal system in schizophrenia a magnetic resonance imaging study. *Archives of General Psychiatry, 43,* 136-144. doi:10.1001/archpsyc.1986.01800020042006

Apple, D. M., Fonseca, R. S., & Kokovay, E. (2017). The role of adult neurogenesis in psychiatric and cognitive disorders. Brain research, 1655, 270-276.

Bauerle, J. D., Grenz, A., Kim, J. H., Lee, H. T., & Eltzschig, H. K. (2011). Adenosine generation and signaling during acute kidney injury. *Journal of the American Society of Nephrology*, *22*(1), 14-20.

Borota, D., Murray, E., Keceli, G., Chang, A., Watabe, J. M., Ly, M., ... & Yassa, M. A. (2014). Post-study caffeine administration enhances memory consolidation in humans. Nature neuroscience, 17(2), 201-203.

Curtis, D. R., & Ryall, R. W. (1966). The excitation of Renshaw cells by cholinometics. *Experimental Brain Research, 2,* 49-65.

Daanen, H. A. (2003). Finger cold-induced vasodilation: a review. *European journal of applied physiology*, *89*(5), 411-426.

Deaglio, S., Dwyer, K. M., Gao, W., Friedman, D., Usheva, A., Erat, A., ... & Kuchroo, V. K. (2007). Adenosine generation catalyzed by CD39 and CD73 expressed on regulatory T cells mediates immune suppression. *Journal of Experimental Medicine*, *204*(6), 1257-1265.

De Bellis, M. D., Keshavan, M. S., Beers, S. R., Hall, J., Frustaci, K., Masalehdan, A., ... Boring, A. M. (2001). Sex differences in brain maturation during childhood and adolescence. *Cerebral Cortex, 11,* 552-557. doi : 10.1093/cercor/11.6.552

Domino, E. F. (2001). Nicotine and tobacco dependence: Normalization or stimulation? *Alcohol, 24,* 83-86.

Domino, E. F., Kadoya, C., & Matsuoka, S. (1994). Recovery cycle of the Hoffmann reflex of tobacco smokers and nonsmokers: Relationship to plasma nicotine and cotinine levels. *European Journal of Clinical Pharmacology, 46,* 527-532.

Domino, E. F., & von Baumgarten, A. M. (1969). Tobacco cigarette smoking and patellar reflex depression. *Clinical Pharmocology and Therapeutics, 10,* 72-79.

Dourado, M., & Sargent, P. B. (2002). Properties of nicotinic receptors underlying Renshaw cell excitation by α-motor neurons in neonatal rat spinal cord. *Journal of Neurophysiology, 87,* 3117-3125. doi:10.1152/jn.00745.2001

Erhart, S. M., Marder, S. R., & Carpenter, W. T. (2006). Treatment of schizophrenia negative symptoms: Future prospects. *Schizophrenia Bulletin, 32,* 234-237. doi:10.1093/schbul/sbj055

Evenson, R. C., Meier, S. T., & Hagan, B. J. (1993). Sex differences in the age of onset of affective disorders. *Comprehensive Psychiatry,34,* 187-191.

Fannon, D., Chitnis, X., Doku, V., Tennakoon, L., O'Ceallaigh, S., Soni, W., … Sharma, T. (2000). Features of structural brain abnormality detected in first-episode psychosis. *The American Journal of Psychiatry, 157,* 1829-1834. doi:10.1176/appi.ajp.157.11.1829

Flouris, A. D., Westwood, D. A., Mekjavic, I. B., & Cheung, S. S. (2008). Effect of body temperature on cold induced vasodilation. *European journal of applied physiology*, *104*, 491-499.

Giniatullin, R., Nistri, A., & Yakel, J. L. (2005). Desensitization of nicotinic Ach receptors: Shaping cholinergic signaling. *Trends in Neurosciences, 28,* 371-378. doi:10.1016/j.tins.2005.04.009

Ginzel, K. H., Estavillo, J., & Eldred, E. (1975). Nicotine-induced reflex depression of α motoneuron activity in the absence of fusimotor-spindle feedback. *Journal of Neuroscience Research, 1,* 253-265. doi: 10.1002/jnr.490010307

Goldstein, J. M., Tsuang, M. T., & Faraone, S. V. (1989). Gender and schizophrenia: Implications for understanding the heterogeneity of the illness. *Psychiatry Research, 28,* 243-253.

Häfner, H. (2003). Gender differences in schizophrenia. *Psychoneuroendocrinology, 28,* 17-54. doi:10.1016/S0306-4530(02)00125-7

Kelly, K. M., Mikell, C. B., & McKhann, G. M. (2014). Morning Joe or After-Dinner Espresso? Improved Memory Consolidation After Caffeine Administration. Neurosurgery, 74(6), N8-N11.

King, K. T., & Ryall, R. W. (1981). A re-evaluation of acetylcholine receptors on feline Renshaw cells. *British Journal of Pharmacology, 73,* 455-460.

Kingma, C. F., Hofman, I. I., & Daanen, H. A. M. (2019). Relation between finger cold-induced vasodilation and re-warming speed after cold exposure. *European Journal of Applied Physiology, 119*, 171-180.

Lee, H., & Thuret, S. (2018). Adult Human Hippocampal Neurogenesis: Controversy and Evidence. Trends in Molecular Medicine.

Lindekleiv HM, Valen-Sendstad K, Morgan MK, Mardal KA, Faulder K, Magnus JH, Romner B, Ingebrigtsen T. Sex differences in intracranial arterial bifurcations. *Gend. Med.* 2010;7(2):149-55.

Molina, V., Sanz, J., Reig, S., Martinez, R., Sarramea, F., Luque, R., ... Desco, M. (2005). Hypofrontality in men with first-episode psychosis. *The British Journal of Psychiatry, 186,* 203-208. doi: 10.1192/bjp.186.3.203

More effects of cigarettes (1969). *The Lancet, 293,* 1013. Retrieved from http://www.sciencedirect.com/science/article/pii/S0140673669918066

Mustafa, S., Thulesius, O., & Ismael, H. N. (2004). Hyperthermia-induced vasoconstriction of the carotid artery, a possible causative factor of heatstroke. *Journal of Applied Physiology, 96*(5), 1875-1878.

Rao, M. L., & Kölsch, H. (2003). Effects of estrogen on brain development and neuroprotection – implications for negative symptoms in schizophrenia. *Psychoneuroendocrinology, 28,*83-96. doi:10.1016/S0306-4530(02)00126-9

Romanovsky, A. A., & Blatteis, C. M. (1996). Heat stroke: opioid-mediated mechanisms. *Journal of applied physiology, 81*(6), 2565-2570.

Salokangas, R. K. R., & Stengard, E. (1990). Gender and short term outcome in schizophrenia. *Schizophrenia Research, 3,* 333-345.

Schey, B. M., Williams, D. Y., & Bucknall, T. (2010). Skin temperature and core-peripheral temperature gradient as markers of hemodynamic status in critically ill patients: a review. *Heart & Lung, 39*(1), 27-40.

Shefner, J. M., Berman, S. A., & Young, R. R. (1993). The effect of nicotine on recurrent inhibition in the spinal cord. *Neurology, 43,* 2647-2651.

Silverstein, B. (1982). Cigarette smoking, nicotine addiction, and relaxation. *Journal of Personality and Social Psychology, 42,* 946-950. doi.10.1037/0022-3514.42.5.946

Troisi, A., Pasini, A., & Spalletta, G. (2001). Season of birth, gender and negative symptoms in schizophrenia. *European Psychiatry, 16,* 342-348.

Vainshenker, Y., Zinserling, V., Korotkov, A., & Medvedev, S. (2017). Noncanonical Adult Human Neurogenesis and Axonal Growth as Possible Structural Basis of Recovery From Traumatic Vegetative State. Clinical Medicine Insights: Case Reports, 10, 1179547617732040.

Ziauddeen, H., Dibben, C., Kipps, C., Hodges, J. R., & McKenna, P. J. (2011). Negative schizophrenic symptoms and the frontal lobe syndrome: One and the same? *European Archives of Psychiatry and Clinical Neuroscience, 261,* 59-67. doi.10.1007/s00406-010-0133-y

Zimny M, Kawlewska E, Hebda A, Wolański W, Ładziński P, Kaspera W. Wall shear stress gradient is independently associated with middle cerebral artery aneurysm development: a case-control CFD patient-specific study based on 77 patients. *BMC Neurol.* 2021;21(1):1-10.

www.ingramcontent.com/pod-product-compliance
Lightning Source LLC
Chambersburg PA
CBHW040911210326
41597CB00029B/5044